~A BINGO BOOK~

Number The[ory] and Graphing Bingo Book

COMPLETE BINGO GAME IN A BOOK

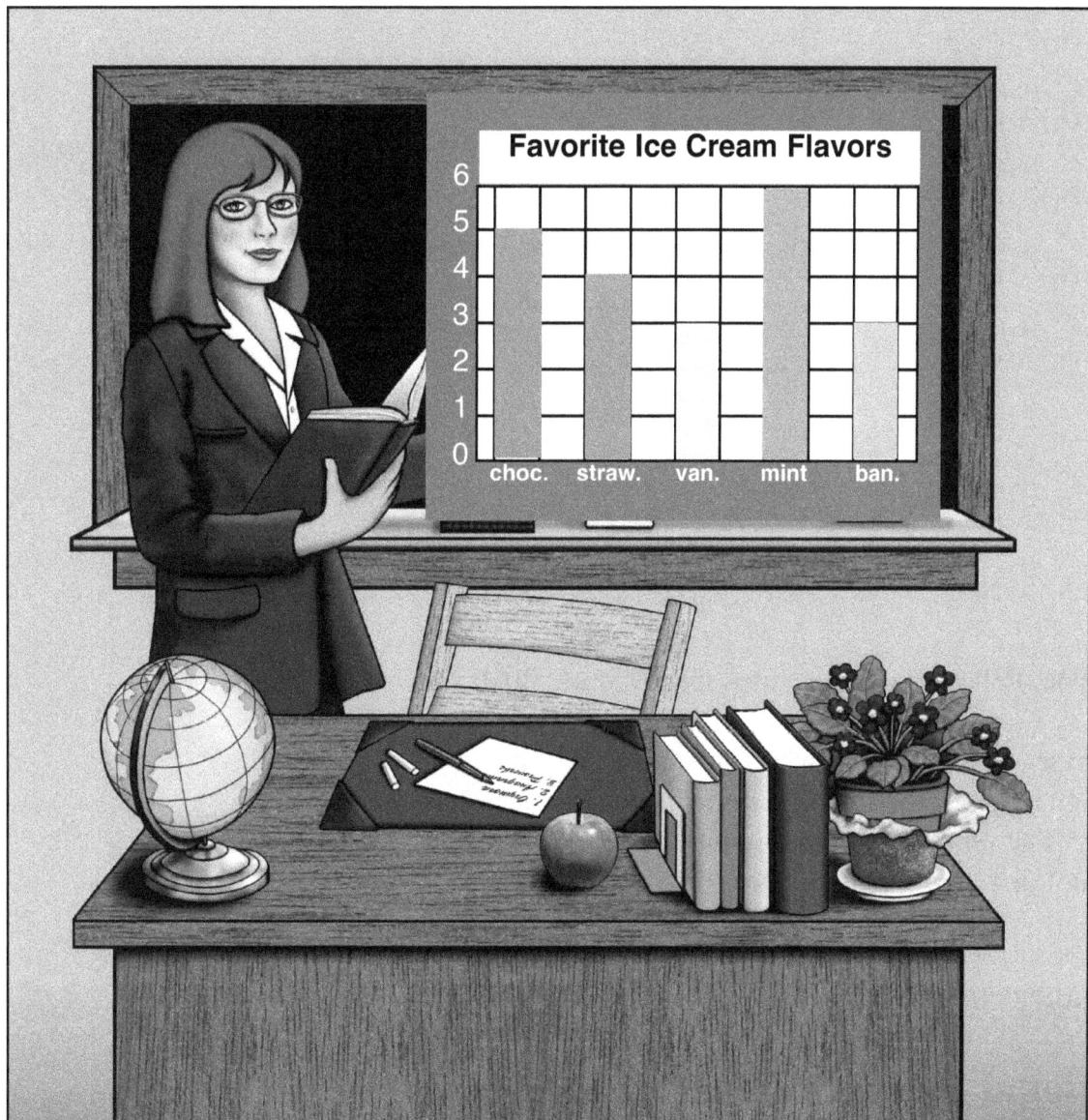

Favorite Ice Cream Flavors

Written By Rebecca Stark

Educational Books 'n' Bingo

TITLE: Number Theory and Graphing Bingo
AUTHOR: Rebecca Stark

ISBN 978-0-87386-452-7

Educational Books 'n' Bingo

Printed in the U.S.A.

NUMBER THEORY AND GRAPHING BINGO DIRECTIONS

INCLUDED:

List of Terms

Templates for Additional Terms and Clues

2 Clues per Term

30 Unique Bingo Cards

Markers

1. **Either cut apart the book or make copies of ALL the sheets. You might want to make an extra copy of the clue sheets to use for introduction and review. Keep the sheets in an envelope for easy reuse.**

2. Cut apart the call cards with terms and clues.

3. Pass out one bingo card per student. There are enough for a class of 30.

4. Pass out markers. You may cut apart the markers included in this book or use any other small items of your choice.

5. Decide whether or not you will require the entire card to be filled. Requiring the entire card to be filled provides a better review. However, if you have a short time to fill, you may prefer to have them do the just the border or some other format. Tell the class before you begin what is required.

6. There are 50 terms. Read the list before you begin. If there are any terms that have not been covered in class, you may want to read to the students the term and clues before you begin.

7. There is a blank space in the middle of each card. You can instruct the students to use it as a free space or you can write in answers to cover terms not included. Of course, in this case you would create your own clues. (Templates provided.)

8. Shuffle the cards and place them in a pile. Two or three clues are provided for each term. If you plan to play the game with the same group more than once, you might want to choose a different clue for each game. If not, you may choose to use more than one clue.

9. Be sure to keep the cards you have used for the present game in a separate pile. When a student calls, "Bingo," he or she will have to verify that the correct answers are on his or her card AND that the markers were placed in response to the proper questions. Pull out the cards that are on the student's card keeping them in the order they were used in the game. Read each clue as it was given and ask the student to identify the correct answer from his or her card.

10. If the student has the correct answers on the card AND has shown that they were marked in response to the *correct questions,* then that student is the winner and the game is over. If the student does not have the correct answers on the card OR he or she marked the answers in response to *the wrong questions,* then the game continues until there is a proper winner.

11. If you want to play again, reshuffle the cards and begin again.

Have fun!

TERMS/ANSWERS

1	DATA
6	DIVISIBLE
7	EVEN NUMBER
9	EXPONENT
11	FACTORS
$\sqrt{144}$	FREQUENCY
13	HISTOGRAM
15	LINE GRAPH
16	LINE PLOT
24	MEAN
25	MEDIAN
27	MODE
32	MULTIPLES
4^3	ODD NUMBER
49	PARALLEL
72	PICTOGRAPH
81	POWER
100	PRIME FACTORIZATION
198	PRIME NUMBER
240	RANGE
BAR GRAPH	SQUARE NUMBER
CIRCLE GRAPH	SQUARE ROOT
COMMON FACTOR(S)	STEM-AND-LEAF PLOT
COMPOSITE NUMBER	SURVEY
COORDINATE(S)	WHOLE NUMBERS

Additional Terms

Choose as many additional terms as you would like and write them in the squares. Repeat each as desired.
Cut out the squares and randomly distribute them to the class.
Instruct the students to place their square on the center space of their card.

Number Theory and Graphing Bingo

Clues for
Additional Terms

Write three clues for each of your additional terms.

_____	_____
1.	1.
2.	2.
3.	3.
_____	_____
1.	1.
2.	2.
3.	3.
_____	_____
1.	1.
2.	2.
3.	3.

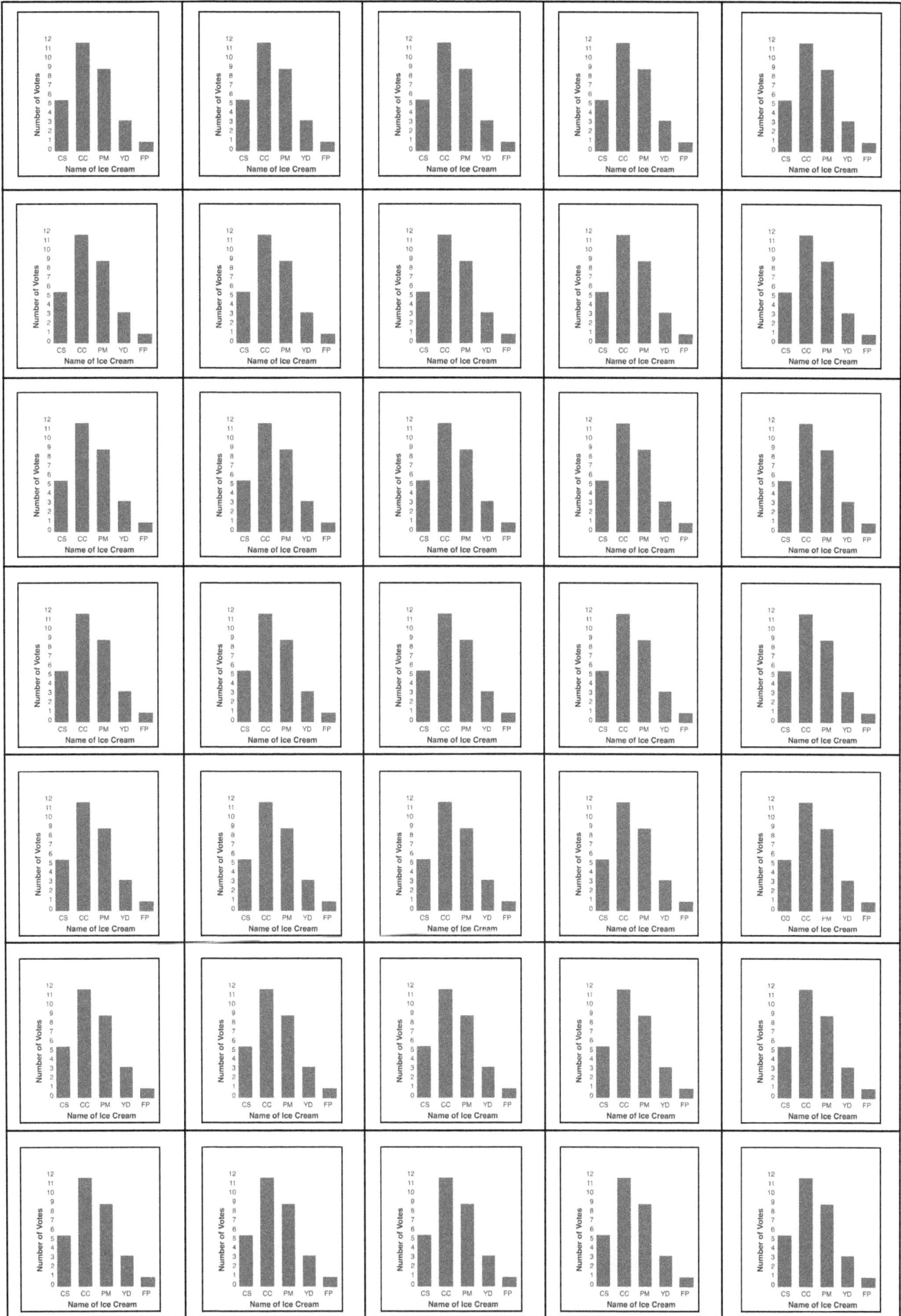

Number of Votes

12
11
10
9
8
7
6
5
4
3
2
1
0

CS   CC   PM   YD   FP

Name of Ice Cream

1 1. When an exponent of a number, the result is always that number. 2. A prime number has two factors: itself and this number. 3. This number is neither prime nor composite.	**6** 1. $\sqrt{36}$ 2. The mean of the following set of data is ___: 4, 5, 5, 6, 8, 8 3. The mode of the following set of data is ___: 5, 6, 6, 6, 7, 8, 11
7 1. $\sqrt{49}$ 2. The range of the following set is ___: 5, 6, 7, 9, 11 and 12. 3. The median of the following set of data is ___: 2, 4, 7, 11, 13	**9** 1. 3^3 2. The lowest common multiple of 3 and 9. 3. $\sqrt{81}$
11 1. It is the only common factor of 44 and 55. 2. $\sqrt{121}$ 3. The mode in this set of data is ___: 10, 10, 11, 11, 11, 11, 12, 13, 14, 14	**$\sqrt{144}$** 1. Square root of 144. 2. 12 3. The equivlalent of 144 ÷ 12.
13 1. Its first six non-zero multiples are 13, 26, 39, 52, 65 and 78. 2. $\sqrt{169}$ 3. The mode of this set is ___: 11, 11, 13, 13, 13, 15, 17, 19	**15** 1. Lowest common multiple of 3 and 5. 2. $\sqrt{225}$ 3. The mean of this set is ___: 10, 13, 14, 15, 16, 17, 20
16 1. Equivalent of 4^2. 2. The mean of the following set of data is ___: 14, 15, 16, 16, 16, 17, 18. 3. The median of the following set is ___: 12, 14, 16, 21, 25	**24** 1. It is the lowest common multiple of 6 and 8. 2. Its factors are 1, 2, 3, 4, 6, 8, 12 and 24. 3. The mode of the following set of data is ___: 18, 18, 24, 24 , 24, 24, 26, 26, 28

Number Theory and Graphing Bingo

25 1. $\sqrt{625}$ 2. Equivalent of 5^2. 3. $5 \times 5 =$___.	**27** 1. It is a common multiple of 3, 9 and 27. 2. Equivalent of 3^3. 3. 3 cubed.
32 1. Equivalent of 2^5. 2. Lowest common multiple of 4 and 16. 3. The range for this set of data is ___: 60, 65, 78, 80, 85, 92	**4^3** 1. 4 to the third power. 2. 4 cubed 3. 64.
49 1. 7 squared 2. The range of this set of data: 20, 25, 37, 41, 55, and 69. 3. The median of this set of data: 41, 42, 42, 50, 55, 56, 57	**72** 1. Its prime factorization is 3 x 3 x 2 x 2 x 2. 2. The exponential form of its prime factorization is $3^2 \times 2^3$. 3. Its factors are 1, 2, 3, 4, 6, 8, 9, 12, 18, 24, 36 and 72.
81 1. Equivalent of 3^4.· 2. $9 \times 9 =$__. 3. Equivalent of 9^2.	**100** 1. Lowest common multiple of 20 and 25. 2. $\sqrt{10,000}$ 3. Its square root is 10.
198 1. Its factors are 1, 2, 3, 9, 11, 18, 22, 66, 99 and 198. 2. Its prime factorization is 3 x 3 x 2 x 11. 3. Its prime factorization in exponential form is 3^2 x 2 x 11.	**240** 1. It is the least common multiple of 12, 16 and 20. 2. $15^2 + 15 =$ ___. 3. The median of this set of data is ___: 160. 210. 240, 250, 280

Number Theory and Graphing Bingo

Bar Graph	**Circle Graph**
1. A graph that uses the height or length of the rectangles to compare data.	1. Also called a pie graph or pie chart.
2. A double one uses two rectangular shapes to show 2 sets of related data.	2. A circular chart that is cut into sections by lines going through its center.
3. It displays data with vertical or horizontal bars.	3. A graph used to show how parts make up a whole.

Common Factor(s)	**Composite Number**
1. When two or more numbers have one or more factors that are the same, we call those factors ___.	1. A whole number with at least one factor besides itself and 1.
2. 8, 16 and 32 have the ___ 1, 2, 4 and 8.	2. A number divisible by at least one number in addition to itself and 1.
3. The greatest ___ of a set of numbers is the largest factor shared by those numbers.	3. The number 2 is the only even number that is not this kind of number.

Coordinate(s)	**Data**
1. In a 2-dimensional plane, they are the pairs of numbers that specify the location of a point. The x-___ is given 1st and is followed by the y-___.	1. Collected pieces of information.
2. When describing the location of a point, the x-___ is given 1st and is followed by the y-___.	2. Sometimes these pieces of information are arranged in a table with rows and columns.
3. The x-___ of a point is its distance from the horizontal axis.	3. Sometimes these pieces of information are arranged by taking a tally.

Divisible	**Even Number**
1. A number is ___ by another if the quotient will be a whole number with no remainder.	1. A number that is divisible by 2.
2. A whole number is ___ by 2 if its last digit has 2 as a factor.	2. It has 0, 2, 4, 6 or 8 in the ones place.
3. A whole number is ___ by 5 if the ones digit is 0 or 5.	3. An even number + an even number results in an ___.

Exponent	**Factors**
1. The small raised number used to show that we want to multiply a factor by itself.	1. Numbers which, when multiplied together, result in a certain product.
2. In 4^2, the small number 2 is one. The number 4 is the base.	2. 1, 2, 3, 4, 6, 8, 12 and 24 are ___ of 24.
3. In 5^3, the small number 3 is one. The number 5 is the base.	3. 1, 2, 3, 4, 5, 6, 9, 12 and 18 are ___ of 36.

Number Theory and Graphing Bingo

Frequency	**Histogram**
1. The number of times a particular data value occurs in a given time.	1. This type of graph uses rectangles to show frequency.
2. Tally marks are helpful in keeping track of this.	2. A graphical display of tabulated frequencies.
3. Histograms are sometimes used to show this.	3. It is similar to a bar graph, but ti shows the number of data items that occur within each interval.

Line Graph	**Line Plot**
1. Shows a change over time with points that are connected by line segments.	1. To make one, you start by drawing a horizontal number line.
2. A double one uses line segments to show two sets of related data over a period of time.	2. Each value in the set is recorded by placing an "*x*" over the corresponding value.
3. To make one, mark the coordinate grid with dots and join the dots to form lines.	3. By looking at the number of *x*'s over the number line on this type of chart, you can see how often the value occurred.

Mean	**Median**
1. The average of all the numbers in a set of data.	1. The middle value in a set of numbers.
2. In the following set of data, the ___ is 5: 4, 4, 4, 6, 7	2. In the following set of numbers, the ___ is 8: 4, 5, 6, 8, 10, 11, 15
3. In the following set of data, the ___ is 14: 10, 10, 15, 15, 20	2. In the following set of numbers, the ___ is 25: 12, 18, 22, 25, 28, 30, 32

Mode	**Multiples**
1. The value that occurs most frequently in a set of numbers.	1. The products of a number and other factors are ___ of that number.
2. In the following set of numbers, the ___ is 2: 1, 2, 2, 2, 2, 3, 4, 5, 5, 6, 7, 7, 8	2. Non-zero ___ of 4 are 4, 8, 12, 16 and so on.
3 In the following set of numbers, the ___ is 8: 4, 5, 5, 6, 7, 7, 8, 8, 8, 8, 9	3. Common ___ of 8 and 12 are 24, 48 and 72.

Odd Number	**Parallel**
1. A number not divisible by 2.	1. This term is used to discribe lines in the same plane that never meet.
2. It has 1, 3, 5, 7 or 9 in the ones place.	2. Lines in a bar graph are this.
3. An odd number + an even number = an ___.	3. Describes the horizontal or vertical bars in a bar graph.

Pictograph	**Power**
1. A graph that uses pictures to show data.	1. Used to describe the result of the multiplication of a number by itself.
2. This simple type of graph uses symbols and/or pictures to compare data.	2. 2 to the second ___ means 2 squared.
3. Sometimes a key is needed to explain the pictures or symbols on this type of graph.	3. 2 to the third ___ means 2 cubed.

Prime Factorization	**Prime Number**
1. The breaking of a composite number into its prime factors.	1. A whole number with only 2 factors: itself and 1.
2. The ___ of 100 is 5 x 5 x 2 x 2.	2. It is divisible only by itself and 1.
3. The ___ of 100 in exponential form is $5^2 \times 2^2$.	3. The first 7 are 2, 3, 5, 7, 11, 13 and 17. The only even number that is a ___ is the number 2.

Range	**Square Number**
1. The difference between the largest number and the smallest in a set of data.	1. This results when a number is multiplied by itself.
2. In the following set of numbers, the ___ is 4: 2, 2, 2, 3, 4, 5, 6	2. We say a number is a perfect ___ when its square root is an integer.
3. In the following set of numbers, the ___ is 6: 1, 2, 4, 4, 5, 5, 7, 7	3. Some examples are 16, 49, 64 and 81.

Square Root	**Stem-and-Leaf Plot**
1. One of two identical factors of a square number.	1. A method of organizing data in a way that shows its shape and distribution.
2. Its symbol is called a radical sign.	2. In a ___ the ones digits and the tens and greater digits are separated by a line.
3. The ___ of 81 is 9.	3. The ones digits are its stems and the tens and greater digits are its leaves.

Survey	**Whole Numbers**
1. The process of asking a group of people the same question.	1. Positive numbers without a fraction or a decimal point.
2. Tally marks are often used to keep track of people's answers when taking one.	2. The numbers 2, 15 and 20 are examples. The numbers -2, 3 1/2 and 6.25 are not.
3. If you want to find our how many siblings your classmates have, you might take one.	3. The numbers 5, 106 and 21 are examples. The numbers -5, 106.5 and 1/5 are not.

Number Theory and Graphing Bingo

Number Theory and Graphing Bingo

Even Number	1	11	Bar Graph	Coordinate(s)
Common Factors	27	Square Number	198	Factors
9	Stem-and-Leaf Plot		Pictograph	Divisible
$\sqrt{144}$	4^3	Survey	Line Plot	Parallel
Multiples	81	240	Frequency	Exponent

© Barbara M. Peller

Number Theory and Graphing Bingo

$\sqrt{144}$	9	Line Graph	Range	32
Parallel	198	15	4^3	Mode
25	81		49	Survey
Odd Number	Prime Number	Stem-and-Leaf Plot	Prime Factorization	Exponent
Factors	Square Number	240	Common Factors	Frequency

Number Theory and Graphing Bingo

$\sqrt{144}$	Survey	198	Line Plot	9
81	6	Range	1	Data
4^3	Square Number		Mode	7
Stem-and-Leaf Plot	25	Multiples	Odd Number	Line Graph
Frequency	Common Factors	240	Prime Factorization	49

© Barbara M. Peller

Number Theory and Graphing Bingo

Stem-and-Leaf Plot	Mode	11	Common Factors	49
Histogram	13	1	Range	9
Pictograph	Odd Number		Coordinate(s)	Bar Graph
Survey	72	Square Number	240	15
Data	Factors	Whole Numbers	Frequency	Divisible

Number Theory and Graphing Bingo

Factors	Coordinate(s)	4^3	15	Common Factors
Histogram	Survey	24	32	6
11	Divisible		Mean	100
Exponent	49	Even Number	Prime Factorization	27
198	240	9	Stem-and-Leaf Plot	Pictograph

Number Theory and Graphing Bingo: Card No. 5

Number Theory and Graphing Bingo

7	Mode	Line Graph	49	Divisible
Line Plot	4^3	27	1	9
Range	Circle Graph		13	32
240	Multiples	Prime Factorization	Whole Numbers	11
Parallel	Survey	Even Number	Pictograph	72

Number Theory and Graphing Bingo

Even Number	Mode	240	Mean	198
Parallel	32	81	16	Histogram
Line Graph	Bar Graph		49	13
Stem-and-Leaf Plot	Odd Number	24	$\sqrt{144}$	25
240	Common Factors	Prime Factorization	Whole Numbers	7

Number Theory and Graphing Bingo

Pictograph	Mode	16	Line Plot	13
Histogram	11	Range	Divisible	15
72	Power		49	Coordinate(s)
Frequency	Stem-and-Leaf Plot	$\sqrt{144}$	Circle Graph	Odd Number
Square Number	240	Whole Numbers	4^3	Parallel

Number Theory and Graphing Bingo

49	198	81	72	Common Factors
Circle Graph	32	Pictograph	4^3	Mode
Data	Even Number		27	16
6	Exponent	Multiples	Mean	100
Odd Number	Prime Factorization	24	$\sqrt{144}$	Coordinate(s)

Number Theory and Graphing Bingo

$\sqrt{144}$	Line Plot	13	Range	72
Divisible	15	1	27	49
Power	Mode		Bar Graph	25
Multiples	Exponent	6	Prime Factorization	Data
24	Parallel	Line Graph	Factors	Pictograph

Number Theory and Graphing Bingo

7	Mode	4^3	27	Parallel
16	Data	Mean	32	1
Histogram	49		Line Graph	81
24	9	Prime Factorization	Common Factors	$\sqrt{144}$
Circle Graph	240	Even Number	Whole Numbers	198

Number Theory and Graphing Bingo

198	Coordinate(s)	13	Line Plot	49
81	Parallel	11	Whole Numbers	27
Even Number	100		Divisible	Range
240	Odd Number	32	$\sqrt{144}$	Histogram
Mode	16	Power	Circle Graph	15

Number Theory and Graphing Bingo

27	Coordinate(s)	7	Factorization	Divisible
11	16	32	49	25
Line Plot	15		81	100
Pictograph	Prime Factorization	13	Power	$\sqrt{144}$
240	Exponent	Whole Numbers	Even Number	Mean

Number Theory and Graphing Bingo

Common Factors	32	4^3	49	Circle Graph
15	Even Number	Data	27	Mode
6	Bar Graph		Line Graph	24
Exponent	Prime Factorization	Power	13	7
240	Range	25	Parallel	Pictograph

Number Theory and Graphing Bingo

Mean	32	4^3	198	Line Plot
7	Line Graph	1	11	Circle Graph
Divisible	Even Number		9	Mode
240	Data	16	Prime Factorization	27
Parallel	Odd Number	Whole Numbers	72	81

Number Theory and Graphing Bingo

13	Data	16	72	Prime Number
Range	25	100	Histogram	Bar Graph
6	Coordinate(s)		Divisible	81
Stem-and-Leaf Plot	15	240	Mean	$\sqrt{144}$
Circle Graph	Square Root	Whole Numbers	Odd Number	Mode

Number Theory and Graphing Bingo

24	Median	Composite Number	Data	Common Factors
Mean	Circle Graph	Prime Factorization	Bar Graph	100
32	Pictograph		Square Root	16
Exponent	Parallel	$\sqrt{144}$	4^3	25
Multiples	6	198	Line Plot	Coordinate(s)

Number Theory and Graphing Bingo

72	Power	15	27	Range
Mode	24	Multiples	Divisible	Circle Graph
32	25		Composite Number	11
Exponent	1	Prime Factorization	$\sqrt{144}$	Line Graph
Square Root	Data	4^3	Median	7

Number Theory and Graphing Bingo

Divisible	7	Data	16	Power
Mean	Line Plot	Mode	198	Bar Graph
Median	Common Factors		27	9
Line Graph	Square Root	Multiples	Odd Number	Composite Number
11	Prime Number	Parallel	Pictograph	Whole Numbers

Number Theory and Graphing Bingo

Power	Median	Line Plot	Data	Whole Numbers
15	81	Histogram	Multiples	Range
Coordinate(s)	100		Stem-and-Leaf Plot	1
Factors	Square Number	Frequency	Odd Number	Square Root
Survey	Pictograph	Prime Number	$\sqrt{144}$	Composite Number

Number Theory and Graphing Bingo

Mean	7	Histogram	Data	Factors
Coordinate(s)	Composite Number	13	16	Even Number
25	Parallel		Median	4^3
Multiples	198	Square Root	Exponent	Pictograph
Stem-and-Leaf Plot	Prime Number	Whole Numbers	24	Odd Number

Number Theory and Graphing Bingo

72	Line Graph	Composite Number	11	27
Range	Line Plot	9	16	6
15	Bar Graph		Even Number	100
Square Root	Exponent	Odd Number	1	Common Factors
Prime Number	24	Median	25	Histogram

Number Theory and Graphing Bingo

13	Median	198	11	Whole Numbers
7	Power	Parallel	Mean	1
Line Graph	27		Frequency	Even Number
25	Prime Number	Square Root	24	Odd Number
Factors	Square Number	Pictograph	Multiples	Composite Number

Number Theory and Graphing Bingo

13	Power	Common Factors	Median	16
Composite Number	Whole Numbers	Histogram	Range	Even Number
100	72		27	25
Factors	Frequency	Square Root	24	Coordinate(s)
Survey	Stem-and-Leaf Plot	Prime Number	Line Plot	Square Number

Number Theory and Graphing Bingo

Stem-and-Leaf Plot	Histogram	Median	4^3	Composite Number
1	Exponent	Mean	13	6
Coordinate(s)	16		Frequency	Square Root
9	Factors	Square Number	Prime Number	Bar Graph
Whole Numbers	Common Factors	15	Circle Graph	Survey

Number Theory and Graphing Bingo

Composite Number	Median	Line Graph	Range	72
Multiples	Line Plot	16	Power	13
Exponent	Frequency		Bar Graph	Stem-and-Leaf Plot
24	11	Factors	Prime Number	Square Root
100	Circle Graph	4^3	Square Number	Survey

Number Theory and Graphing Bingo

Line Graph	15	Median	Power	81
Factors	Frequency	Mean	Square Root	6
Prime Factorization	Square Number		Prime Number	Stem-and-Leaf Plot
72	7	Histogram	Survey	1
Circle Graph	Bar Graph	Composite Number	9	100

Number Theory and Graphing Bingo

Divisible	Power	9	Median	13
81	Composite Number	Frequency	Range	Bar Graph
Square Number	25		100	Multiples
$\sqrt{144}$	72	Parallel	Prime Number	Square Root
11	32	Circle Graph	Survey	Factors

Number Theory and Graphing Bingo

Composite Number	Power	72	Mean	32
Exponent	Multiples	Histogram	100	9
Coordinate(s)	Frequency		6	Median
81	Factors	49	Prime Number	Square Root
13	6	Survey	7	Square Number

Number Theory and Graphing Bingo

Common Factors	Median	Range	49	Square Root
1	Power	Line Graph	Bar Graph	27
Exponent	6		100	Histogram
Survey	7	11	Prime Number	Frequency
Factors	198	Square Number	Composite Number	9

Number Theory and Graphing Bingo: Card No. 30